Gerardo Sánchez

The Secrets of Quail Breeding

Gerardo Sánchez

The Secrets of Quail Breeding

Secrets

ScienciaScripts

Imprint

Any brand names and product names mentioned in this book are subject to trademark, brand or patent protection and are trademarks or registered trademarks of their respective holders. The use of brand names, product names, common names, trade names, product descriptions etc. even without a particular marking in this work is in no way to be construed to mean that such names may be regarded as unrestricted in respect of trademark and brand protection legislation and could thus be used by anyone.

Cover image: www.ingimage.com

This book is a translation from the original published under ISBN 978-620-0-01283-8.

Publisher:
Sciencia Scripts
is a trademark of
Dodo Books Indian Ocean Ltd., member of the OmniScriptum S.R.L Publishing group
str. A.Russo 15, of. 61, Chisinau-2068, Republic of Moldova Europe
Printed at: see last page
ISBN: 978-620-4-09728-2

Los Secretos de la Cría

de

Codorniz

Por Gerardo Sánchez

The Secrets of Quail Breeding

INTRODUCTION

This book has been written with the sole purpose of providing knowledge for those professional and amateur breeders around the world who are passionate about the amazing world of quail breeding. In the same way, the material will provide the reader with a lot of important data that will be necessary at the time of expanding your breeding or simply start as a breeder. All this has been elaborated mostly thanks to my experience in the area and, of course, to the review of specialized technical materials as a means of support.

The coturnix japonica or quail is an excessively profitable bird, so much so that every day there are more countries and people who are integrated into this wonderful activity. This type of bird comes from Japan and is classified as gallinaceous, thus, (***Lucotte G.***) Pag.7 sustains that ¨...it was introduced in the USA in 1955, a little later in Italy and then in all Europe...¨ The physiological advantages of this animal are considered, among others: the precocity of its laying, its high percentage of fecundity, its fast growth and its great resistance to diseases.

From all aspects, the quail represents an enormous economic wealth, the capital investment is not high and leaves a lot of profitability. If we look at it from the conversion and compared to other larger birds, it is extremely economical because we do not need large amounts of food for maintenance, also, their eggs offer us a variety of benefits such as low percentage in cholesterol levels. Finally, we can mention the exquisiteness of its meat and the nutrients provided in the various dishes offered by many restaurants in the world.

Registration Number: 712145329 - 11326692
Official Timestamp: 2021-08-16 23:02:39 UTC *All rights reserved.*

Gerardo Sanchez

LIST OF CONTENTS

Unit 1

FUNDAMENTAL ASPECTS

The quail, animal well known by the hunters and the gourmets, belongs to the group of the gallinaceous of the genus **Coturnix**, and forms, together with other genera, the group of the quails of the old world. The genus *Coturnix* is, since long, the richest in species and these can be divided in three big groups according to their origin, constituting respectively the African, Asian, Australian and New Guinean groups. The most common species is the *Coturnix coturnix*, which is spread in Europe, Asia, Africa and in the Atlantic islands.

The **Coturnix** is a wild animal, it is the one we can commonly find and it is from our regions, it nests in Europe and Asia and migrates during the winter to Africa, Arabia and India. The other subspecies *Coturnix coturnix coturnix* japonica, is the Japanese quail; it nests in the island of *Sakhaline* and in the archipelago of Japan and migrates to Siam, Indochina and Formosa. This second species was the one that was domesticated long ago in Japan and imported to Europe and the United States.

Domesticated and wild quail are easily recognized by their structure, by the song of the males which is very different in both breeds and by the details of the plumage: in the male, the colour of the breast and chin are much more constant in the domestic breed than in the wild, while in the females the feathers of the same region are spotted black in the domestic quail and pale in the wild.

The domestic quail is a small bird with a weight of approximately 150 g the female and 120 g the male, collected on itself and rounded shapes. The quail chicken during its birth is tiny and its weight is 10 g in its majority, here in Venezuela it is called ¨Cotupollo¨.

It has a striped or mottled down with black bands and a very fast growth. The quail egg is ovoid in shape and can measure 3 cm, its width just under 2.5 cm. The colour and pattern of the egg vary greatly from one layer to another. (Pigmentation of the egg).

It must be recognized that the domestic quail has some particularities that make it superior in poultry farming to any other known gallinacea; the embryonic development is approximately 15 to 16 days, therefore, it is extremely fast, the laying is very early and the individuals are adults from the age of five weeks. Under special lighting conditions the laying percentage is 80 per 100, i.e. approximately 300 eggs per year for each layer on average. A female quail lays almost three kilos of eggs per year, 25 times her own weight, double the production of a laying hen. By rearing one male with two or three females per cage, approximately 80% of the eggs can be counted as fertile, so that five generations per year are possible. Before the age of 45 days, a quail is edible since its weight is 120 g and it has not consumed more than 500 grams of feed.

ILLUSTRATIONS

1- Characteristics of the adult domestic quail, the female on the left has a black breast with black spots while the male on the right has a tan breast.

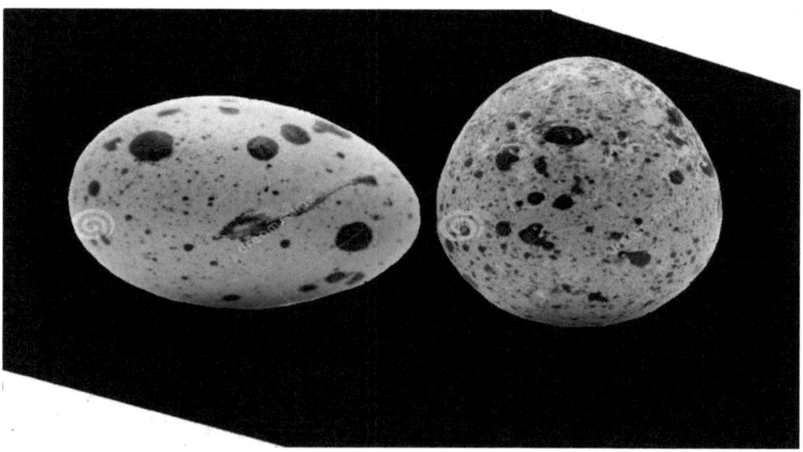

2- Appearance of the eggs, they are spotted (pigmentation) and their color is a distinctive feature of each female.

Unit 2

INSTALLATION OF THE

INSTALLATIONS

About the facilities

It is advisable to place the quails in very large spaces where there is good lighting and adequate ventilation system, it should be noted that this type of bird does not really require a lot of space unlike other specimens such as chickens, broilers, turkeys, etc..

There are different ways to organize when starting production, if you want to start big and you have enough animals, then I advise you to build a shed where you can place a good amount of cages in the form of batteries, this way of organization is much more effective, this way you gain enough space. Also, each battery can contain 5 to 7 special cages for layers.

The other arrangement of cages that I use in particular is the linear or horizontal form, they are tied to the roof beams and at each end of the cages, this system has been very effective when it comes to placing the feed, collecting eggs and cleaning the waste. If we compare the work with the battery cages and the linear cages there is a big difference, the first one is the egg collection, while the person who collects the eggs in the battery has to do it cage by cage, from top to bottom or vice versa, the one who does it in the horizontal cages is much more efficient and faster, this applies to the placement of feed and water. On the other hand, if you look at the cleaning part, the waste falls directly into the channels under the cages; it is worth mentioning that it does not include waste trays like the ones used in the batteries, all the waste goes to a pit or in my case to channels that go directly to the sewage sewers.

As can be seen in the images below, there are advantages and disadvantages in terms of organization, the batteries allow us to take advantage of the establishment because it does not occupy space, while in the cages placed horizontally reduces it.

3- Cages for quails arranged in the form of batteries where the shed is used to the maximum.

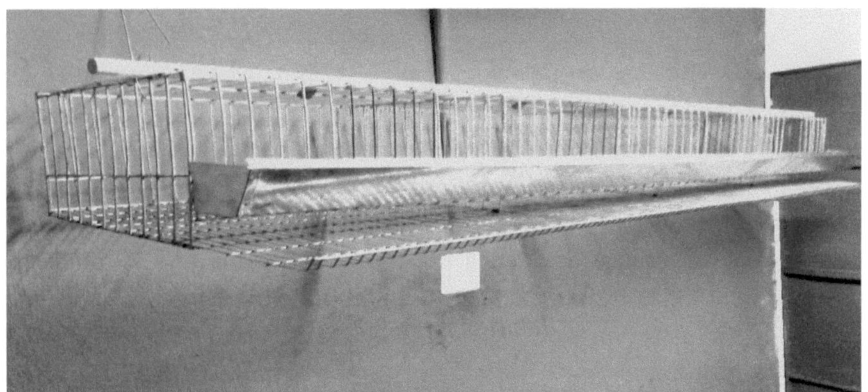

4- Quail cages arranged in lines or horizontally, note that they are fully hung from the ceiling and the floor does not use trays. The droppings go directly into the carcass.

Note that the cages placed in this way allow us to have a lower number of quail per square meter than the battery system shown in picture number 3.

This picture shows the correct way to build a shed for quail breeding.

Air circulation and environment.

The *Coturnix* or *japonica quail* is a bird extremely sensitive to atmospheric changes, therefore, it is recommended that the ideal climate is between 20 degrees in winter and should be avoided in them any sudden change in temperatures. On the other hand, when it is time for the Cotupollos to hatch, it is necessary to have a room ready especially for them, the temperature changes during incubation and at the time of removing them from the incubator should be kept almost similar, that is, in the place where the newborn birds will remain, there should be an environment similar to the one they have just left. If, on the contrary, it lacks a good conditioning that generates enough heat, they could be affected during the first week, resulting in a considerable number of mortality.

The Cotupollo is an extremely delicate bird in this sense, the opposite case to the above, is that if heat is generated in abundance, they themselves will seek to move away to not suffocate, so there must be a balance in temperature, neither too cold nor too hot, oddly enough, the same chicks will teach us when they feel comfortable in a place, just observe their behavior daily

In the case of adult quails, the facilities must be conditioned, there must be good lighting and adequate air flow. V. Rodrigo & B. Hugo 2007 (p. 21) ¨Inside the house, the ideal temperature ranges from 13 to 23 ºC. Free circulation of air must be allowed and ventilation is controlled by curtains. The main function of ventilation is to remove ammonia gases and control water vapour (relative humidity), to help keep the temperature within tolerable limits for the bird. ¨

As can be seen, these authors point out that air currents must be dosed in the poultry houses since this will allow a greater evacuation of all the ammonia produced by the excrement of the birds and at the same time this control will also help us to prevent possible diseases that may arise.

From the right place for the Cotupollos

Returning to the subject of hatching, it can be noted that as far as possible, there should be total asepsis on the part of the farm manager or staff. When the chicks are taken out of the incubator, a sufficiently lighted and spacious room should be prepared where they will be placed; there should be no power supply failures, especially in very cold regions, since the bird is not able to withstand the absence of heat during the initial stage. In this sense, if you do not have a stable service in energy, the most advisable is that you should take precautions such as: to get a generator to maintain lighting.

In cases where there is no generator and there is a critical moment of absence of high temperature, the chicks will immediately seek to crowd together for warmth, this will result in a high percentage of mortality. To avoid this bad experience I will recommend the following:

a) If you are going to have a room, try to close all access to the corners, if you do not, at the time of power failure all the quails will seek to group in one of the corners and die of asphyxiation.
b) Look for alternative means of generating electricity on a continuous basis to maintain the incubations and especially the heat in the Cotupollos.

c) Build a circular room or, if your kennel is small, build bells where they can stay safe and warm.

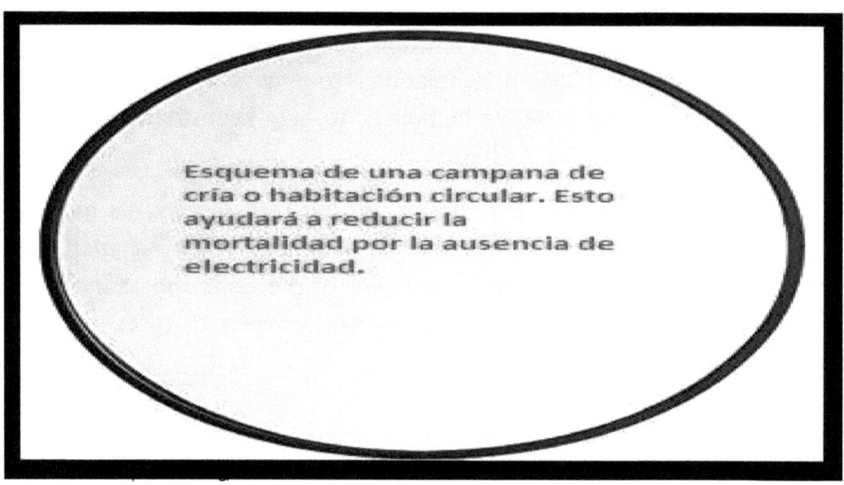

6- Circular room with lighting system for Cotupollo breeding.

The basis of good breeding lies in the construction of appropriate facilities, this simple secret hardly anyone knows and certainly no breeder will tell you. With this system I have been able to reduce to 95% the mortalities caused by the deficiency in the electricity supply. Following these simple instructions will help you in your Cotupollo breeding process.

Cages for quails

There are different types of cages such as the handmade ones that are created by people and the others that are generally used for large-scale quail breeding. To start a breeding farm and obtain an optimal space to work it is advisable to use the cage system manufactured by the companies, its name is the "Roll **Away"** the same are built in wire rods with electro welding point, with a spacing of 2.5 cm between wires so that the birds can feed properly, and likewise the floor is constructed of the same material leaving a small opening of 10 mm that allows the passage of excreta to the tray or carcasses. This type of cage comes with the egg collecting tray, in some cases with about 5 degrees of inclination to allow a smooth displacement of the egg.

Jaula doble Roll Away con capacidad para 50

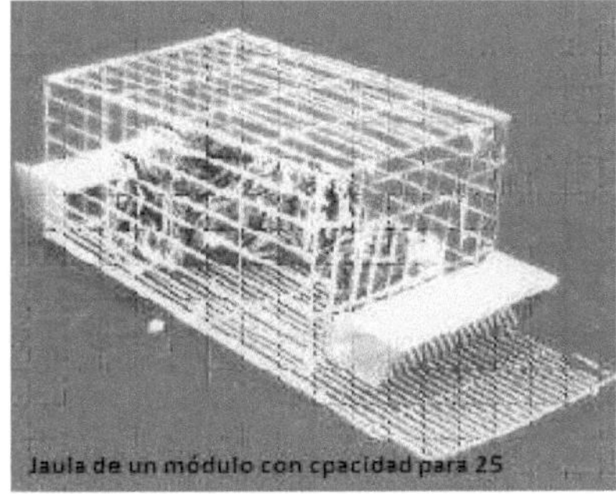

Jaula de un módulo con cpacidad para 25

7- The image shows the type of cage used for egg production.

The cages manufactured in the companies already come with a standard, generally a cage can measure 100 cm long, 50 cm wide by 25 cm high and has a capacity to house between 20 and 25 adult birds per module. In this way, you can make the most of the capacity to form batteries and thus be able to put more animals per square meter. V. Rodrigo & B. Hugo 2007 (p. 24) ¨The batteries are made up of 15 stacked cages, forming 5 floors of 3 cages each, so that the batteries house between 225 and 300 quails, depending on the type of breed used ¨.

These data provided are very useful when placing the number of birds needed in each module, I personally never put more than 20 quail in each division because they need space, it is always recommended that they are unburdened, especially if it is animals for reproduction. The cages in addition to having the egg collection tray, some are already designed with a support inside the floor which serves as a support for the droppings collection tray in case you want to use it to form batteries, thus preventing the feces from the upper cages fall on the lower cages. The entire cage is designed so that the feed and water supply does not come into contact with quail droppings or quail droppings.

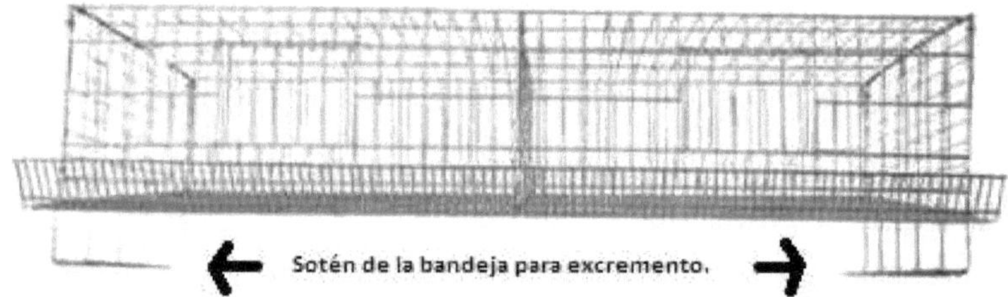

8- Lower part of the quail cage, support of the droppings collection tray.

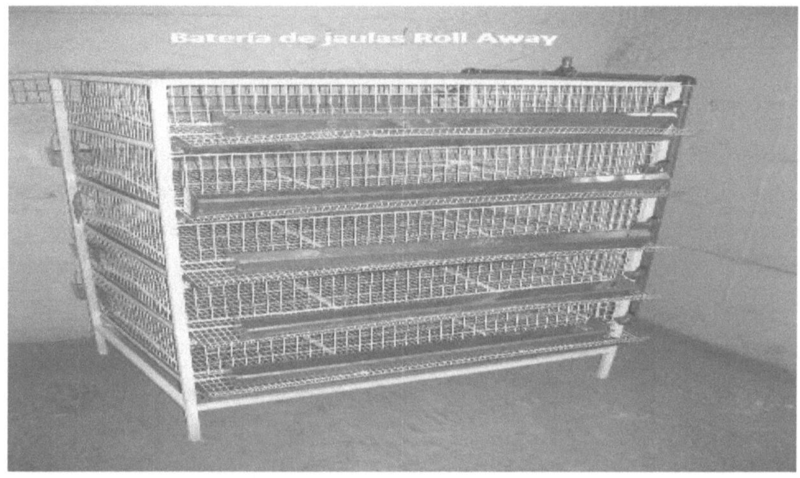

9- Cages arranged in the form of batteries.

From the drinking fountains

The drinking troughs are generally made of plastic as this material allows the elimination of corrosion generated by the water when metal drinking troughs are used, the latter I do not recommend at all.

Types of drinking fountains

The most used in small farms or family farms are linear or grooved and we can place them along the cage at the back of it. The other drinking system used in adult quail is the dripper type where the bird places its beak right on the drop of water coming out of a valve and ingests what is necessary to stay hydrated. This type of drinker is fed by a system of hoses connected to a tank where the water comes down by gravity, the bird touches the small valve and automatically releases the liquid. The advantages of this system is that the water is kept clean. Finally, if it is Cotupollos, the drinkers used are the cup type, this allows the baby can not get into the water but rather ingest what is necessary, it should be noted that this type of drinker is one of the safest that exist for chicks.

From the troughs

There are many varieties of feeders as well as drinkers, but one of the most used are the linear or channel feeders, they are made of aluminum, stainless steel sheets, zinc or plastic, its dimensions vary depending on the age of the animals. If it is at a macro level, the feed dispensers called hoppers are what are responsible for supplying each shed through mechanized systems that supply feed in a controlled manner.

Teams	Fluted feeder 32 birds / linear meter. Fluted drinker 32 birds / linear meter. Drinker for 6 birds.
Density	60 / 64 birds / square meters on each floor. From 10 to 16 birds per compartment.

Table taken from the book Manejo Empresarial del Campo pg. 28

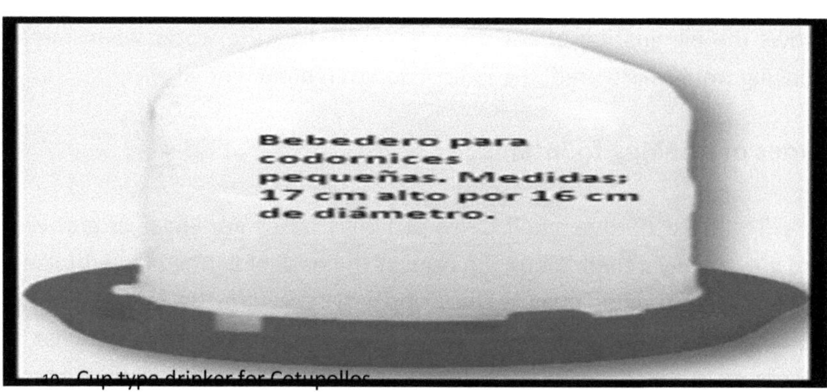

10- Cup type drinker for Cotupollos

11- Automatic drip drinkers.

12- Plastic drinking troughs can also be used as feeding troughs instead of the conventional metal ones.

13- Feeder for adult quails.

14- Feeder for Cotupollos.

Quail: baby quail, usually up to 3 weeks old.

Unit 3

QUAIL NUTRITION

It is of utmost importance a good feeding in birds as this will depend on obtaining a good result when collecting the eggs and the rapid growth of the Cotupollos. The Coturnix or japonica quail requires special food for its development, as I said before, during the initial stage, the chicks are very susceptible to temperature changes but another factor of incidence and very delicate in them is the feeding, surely it is the most critical moment because not any type of food serves. Those who start for the first time in the world of quail breeding must know very well the food requirements; for the newly hatched quail, the key food for its rapid development is the "vitaminized starter", a Cotupollo at the time of birth the only thing it requires is plenty of fat because what we want is growth and maintenance until adulthood. The type of food that I use is ¨Souto Vitaminado starter¨ surely in other countries that company is producing this product, I highly recommend it.

For commercial adult laying quail, a lot of fibre is required in the feed and not so much fat, otherwise if we are preparing it for fattening, at that point, we would already have quail for human consumption. G. Lucotte 1985 (p. 29) states the following: "The nutritional needs are different for quail chickens, broiler quail and breeders...¨ This shows the different nutritional needs for each type of bird. Many people make the mistake of giving the same type of feed for all quail, which is not advisable.

> ...In the case of quail chickens, the ration must cover growth
> and maintenance needs; in the case of broiler quail, it must
> cover supplementary weight gain and maintenance; finally, in
> the case of breeders, it must cover reproduction and laying
> needs, as well as maintenance needs. (Page 29)

In all three parts, the energy content of the feed will depend a lot on the composition, but mainly the chemicals used and especially the antibiotic used in the feed in the case of chicks. G. Lucotte 1985 (p. 30) states that there are three types of commercially available feeds for breeders ¨...It will be of great interest to the breeder to use commercial feeds. There are three kinds of feeds in the trade, feed for the quail chick, feed for the broiler quail and feed for the breeders.¨.

You must have a criterion when making the change of food at different stages of the bird, during the first 4 weeks the quails are consuming the vitaminized starter, I recommend making the transition from food (starter to layer) from 4 weeks, you should be mixing both foods every day with more proportion to the layer. In this sense, it is necessary to emphasize that once the quails are in the laying stage, you should not change the food that is being supplied by another brand, it must be continuous until the end of its laying cycle, if you abruptly interrupt the same type of food that is being given from the 5th week by another, they will automatically stop laying for about 15 days to adapt to the new food. This happens because the components in the food are different, even if you are going to make the change of food in laying quails for any reason, you must take into account that this change should be gradual, not direct.

> The transition from chicken feed to fattening quail feed should be done gradually over several days, passing from two parts chicken feed to one part fattening feed, then one part chicken feed to two parts fattening feed, and finally only fattening feed. During the thirty days of fattening, the quail must be literally satiated with feed in order to reach its weight as quickly as possible. (Page 31)

As you can read, the transition to another feed should be gradual, there should never be a total change unless it is a fattening bird or simply from the stage of baby quails to laying, this also applies to laying birds, if you do not feel comfortable with a certain feed or simply do not get a certain brand in the market for some reason and want your quails do not stop laying, then you should have a good stock of feed you are using and start the change to the new one in a gradual way.

The "ponedora" feed used regularly here in Venezuela for quail breeding is powdered feed from Alimentos La Caridad, also Protinal quail feed or Souto powdered ponedora. There are other brands but due to the problems of the companies to produce the feed caused by the lack of foreign currency and the economic deterioration, these products have become scarce, many farms have had to stop operating permanently.

Below is a table with the necessary requirements in the nutrition of birds, this table is taken as a reference from a research done by G. Lucotte in his book on quail breeding.

	Growth	Fattening	Reproduction
Metabolizable energy, calories / kg -----------------------------------	2.820	2.820	2.800
Crude protein, % ---------------	28,1	24	22,1
Fat, % --------------	3,4	3,2	3,2
Cellulose, % -----------------------	4,1	4,1	3,5
Assimilable phosphorus, % -----------	0,67	0,50	0,44
Calcio -----------------------------	1,26	1,03	2,10

Table on average nutritional requirements for quail chickens, broiler quails and layers.

An adult quail can consume a total of 22 grams of food per day, as you can see, the Coturnix Coturnix or japonica quail is an extremely economical bird in its maintenance, for this reason, many people are attracted to start a breeding farm for commercial purposes, it represents a great business because it leaves a lot of profit margin.

Feed: dry feed given to animals.

Unit 4

COTURNIX COTURNIX

TAXONOMY

Much is said about the classification of the quail, the truth is that it is not easy to recognize the male and female, even among people with experience dedicated to the breeding of this animal. To begin with, the first thing to say is that there are different types of sexing; by the color of the plumage, by the wing and the cloaca, these are basically the most used by breeders.

Sexing by feather colour.

This method of classification of the animal occurs at a glance because the male in the breast has red cinnamon feathers in its entirety, this is accompanied by the chins that are almost always the same tone, its weight is approximately 100 grams and at the top of the cloaca have a reddish seminal gland that when squeezed immediately release a kind of white foam, is the semen. The deep singing of the male is another characteristic feature as well as its sexual hyperactivity. However, the female is usually a little more docile, its weight is about 120 grams and the plumage of the chest is completely gray with black pinticas, unlike the male, she does not have the cinnamon red plumage and his song is imperceptible.

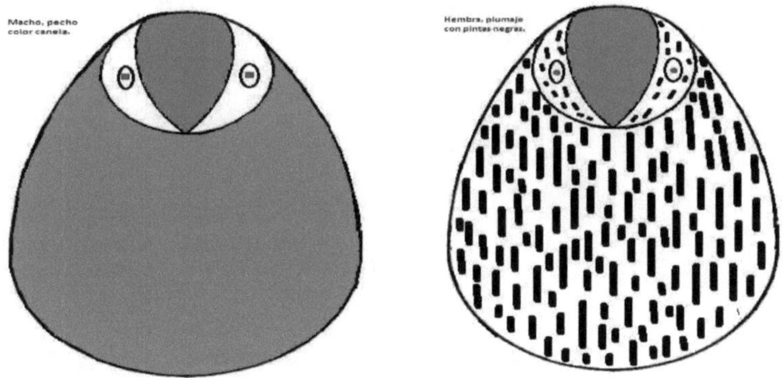

15- On the left, the male with red breast and the female on the right with black spots.

Wing sexing.

This way of sexing the bird is one of the most practical with only 95% error when classifying them, all you have to do is to take a wing by the tip and extend it, then proceed to compare the parity of the down on the back of the wing. In this sense, the females at birth have the feathers of the upper part shorter than those of the lower part, that is to say, that they form a kind of double line in the form of stairs remaining this way without varying until the adult stage. In the opposite case we have the male, this one maintains the same length of the downs from its birth until its adulthood or simply the superior downs surpass the inferior ones, that is to say that the superior downs are not arranged in the form of stairs but the superior ones are of the same length that the inferior ones or surpass them.

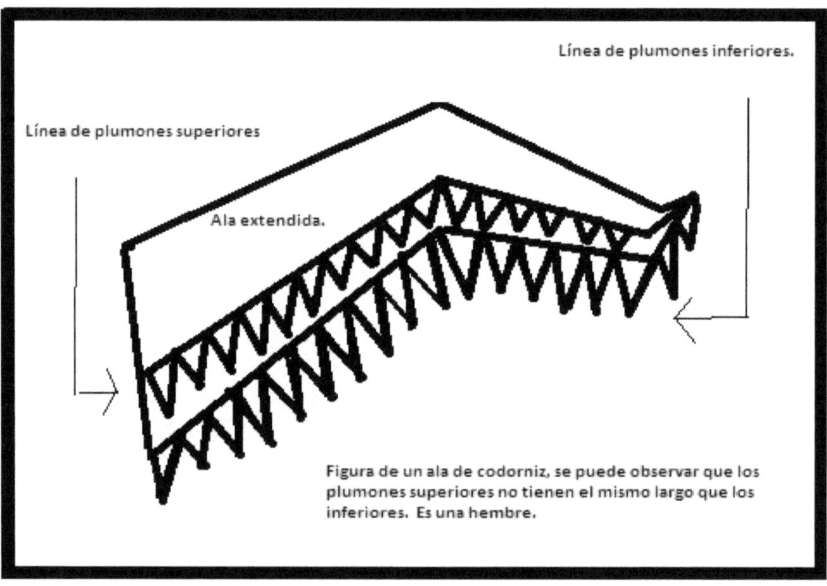

16- Extended wing of the female quail.

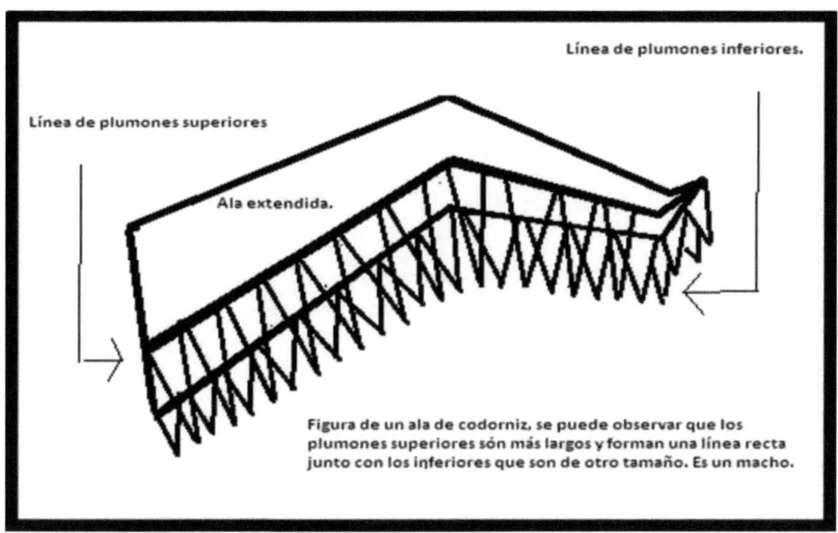

17- Extended wing of a male quail.

Sexing through the sewer.

This type of classification is a little more complicated as it requires a lot of visual acuity and being an expert, not everyone can perform this procedure. Most breeders are not very keen to divulge it as many of them seek to maintain secrecy about this way of classification. Below are some figures with the characteristics of the male and female that are of great importance.

In the case of newborn females, when checking the cloaca, you can notice a kind of "U" in their genitals while the male has a kind of "Y" or beak in their genitals.

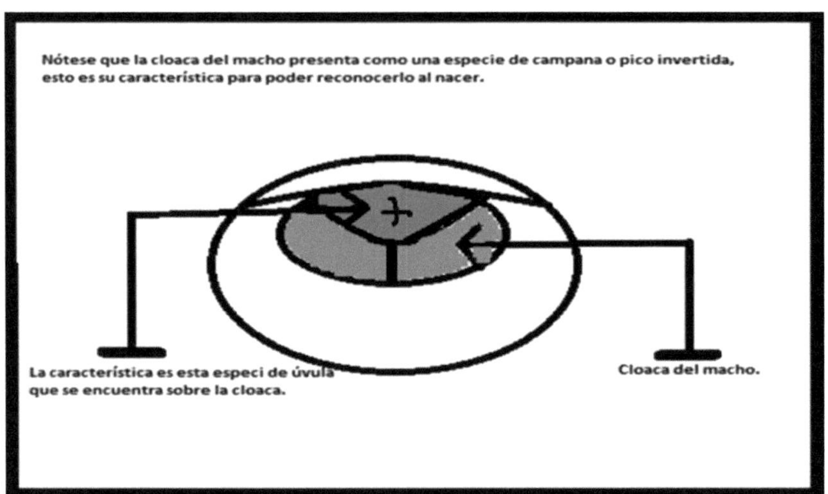

18- Male quail.

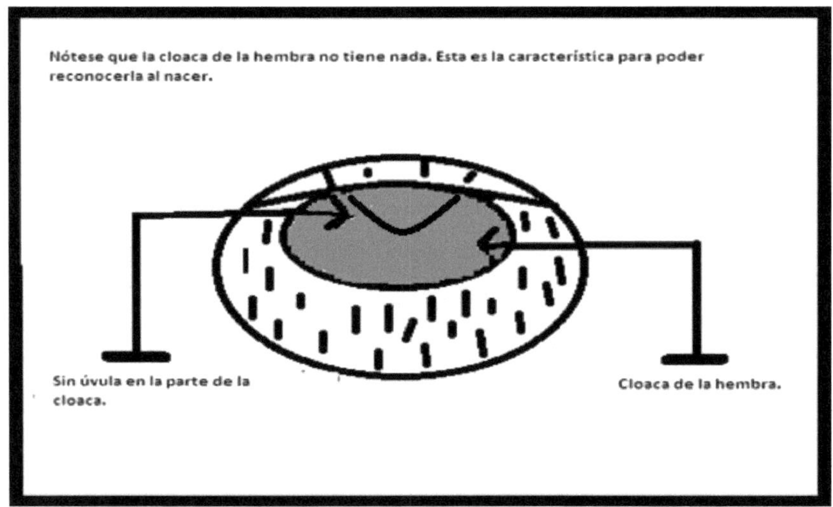

19- Female quail.

Unit 5

ON REPRODUCTION

Breeding and selection of breeding stock.

For the achievement of a good breeding foot many factors are required, one of the most important is the type of male or padrote to use, this must be a bird of good physical condition. On the other hand, feeding is extremely important because for mating requires the energy of the male, one without a good diet will not guarantee the fertility of the eggs at the time of incubation. In many cases the quails are selected by breeders who go so far as to mark them by age, weight and condition by means of a pedigree. In the following G. Lucotte describes the procedure for the selection of breeding birds.

> The success of a breeding depends, in principle, on the quality of the breeding stock that must be acquired from a breeder specialized in the production of high yielding strains. The production of strains needs, in fact, a particular installation and an intense work; the ideal conditions are realized when the breeder does pedigree breeding, and when each genealogy is followed individually, which implies the placement of rings on adults and chicks, as well as the marking of the eggs... (p. 37).

Through this method it can be determined that a large part of the breeders constantly monitor the birds to ensure a good breeding, this will result in a high percentage of fecundity in the eggs and a high level of birth.

Crossbreeding between the same blood.

Mating between the same family of birds should be avoided by all possible means, this almost always leads to malformations or genetic degenerations; quails with crooked beaks, without eyes, with crooked legs, etc. Many times, because of not having an adequate control in the adult modules, a crossbreeding between parents and offspring or between siblings is created, resulting in quails with problems. The general recommendations to avoid this problem are to control the birth rate and family groups. If this is not done, the risk of a high percentage of mortality during embryonic development, hatching and a decrease in egg production increases.

The behaviour of the male.

The male is a highly territorial animal and very aggressive towards his peers of the same sex, and is also sexually active, mating many times with the same female or with a certain number of females in his environment. G. Lucotte 1985 (p.38) states that"...when several males are bred together, a hierarchy is established between them. The dominant male has the priority over the dominated ones for all advantages; feeding, drinking, mating with females...¨".

At the moment of the combats between them for the territorial dominion the wounds are usually inflicted on the eyes, the genital part, the head, and even the loss of the feathers, taking them until the death in many cases. On the other hand, at the moment of the mating the male is accustomed to hold the female by the head using his beak and climbing on his back, immediately afterwards, he approaches his cloaca to the female's, all this happens in a few seconds, the dominant male will catch the first female he has at his disposal and will force her to let him mount her.

From the level of fertility and organization in the modules.

The mating capacity of each male is inexhaustible, usually to obtain a good result in the hatching each of them supports a maximum of 5 females, with this amount the percentage of fecundity is relatively high while if you decrease the number of females per male, this percentage increases to almost 100%. Many breeders advise the use of 3 females per male and it is not criticized since what they want is to obtain a high level of birth, but the wear that the females suffer caused by the sexual activity of the male is considerable.

In this sense, it is possible to develop different ways of organization in the modules to obtain a good result in the birth. The first case is the following:

a) Individual **modules**: this is the most common way in hatcheries of small quantities and non-commercial use, its feature is simple, a male and a female determines the high level of fecundity, the male is placed in the morning and removed after about 30 minutes every day, this also helps to verify if it is a good stud or if on the contrary should be replaced by a younger one.

b) **Collective modules**. In each module of the two that make up a *roll away* cage, a maximum of 30 birds can be housed, that is to say, 20 females are introduced with 10 males in order to guarantee a regular percentage of fecundity. This is used in high level hatcheries since the number of animals is higher than in small hatcheries and the individualization of the couples is practically impossible. The negative effects of this type of collective procedure are the high probability of rivalry between the males.

c) **Small group module**. This procedure is used with a number of 3 to 5 females per male in a small space of the module, this avoids fights between males and raises the fecundity level more in comparison with the module used in a collective way.

Other features in the players.

For the selection of a good breeder it is necessary to highlight the following characteristics that according to Vasquez and Ballesters 2007 (p. 34) are described below:

Males: strong build and well proportioned, lively, with full plumage and in good condition. Dark feathers and cinnamon chest. As intense as possible. Black beak, genital apparatus with a reddish protuberance and the size of a chickpea.

Females: well proportioned and with dark, full and bright plumage. Elongated neck and small head.

Breeders should, if possible, be renewed every year.

Life Cycle

This is the period between the birth of the quail and the end of its egg production and consists of three stages:

- Breeding from 0 to 3 weeks of age; at this stage the management of the reproductive stage is definitive.
- Rising: 4 to 7 weeks of age.
- Posture: from 8 to 60 weeks of age.

In order to obtain a good result in the production of eggs and quail, we must take into account, among other things, the ambient temperature, adequate space, good feed, the storage time of the eggs prior to incubation, the physical conditions of the broodstock, etc. All these details make it possible to efficiently develop a good breeding flock.

Of the birth of the cotupollos.

During the process of collecting fertile eggs it is important to take into account the formation of the eggs and above all their storage period. The incubation period is 16 days from the day of laying in the incubator, with hatching starting in many cases from the day before the scheduled date. This whole process takes place under normal conditions, i.e. the collection time of the fertile eggs shouldnot exceed the stipulated number of days.

Time of collection of fertile eggs.

The ideal collection time to put them in the incubator is 5 days maximum, it is not recommended to exceed this time.

About incubation

The incubation most used by amateur and professional breeders is done by incubators, it is an artificial incubation system and varies in the amount of eggs used by its capacity. Incubators are mostlyelectric, but gas or kerosene incubators are also used. The main factor to highlight is that the incubators come to replace the bird during the entire gestation process, the artificial mother or incubator must maintain all internal conditions such as: adequate humidity level, temperature (37º to 38º) and ventilation for a good embryonic development.

> Temperature: in vertical incubators the temperature should be 37.5º and 38º. The temperature in horizontal incubators should be slightly higher, up to 39º due to heat loss.

> Humidity: in both cases the humidity should be 40 to 50 per 100.

> Ventilation: the vertical incubator and the horizontal incubator are also differentiated by the ventilation mode, while in the first type the air circulation is carried out through ventilation holes, in the second type the air is mechanically agitated inside the apparatus and it is then dynamic ventilation. Lucotte G. (p. 55/56)

The eggs inside the incubator.

During the collection process and before incubation, the quail eggs should be carefully spread out on the tray as they are very fragile and should be checked for cracks in the shell to avoid contaminating the incubator. This often happens due to mishandling or simply because of the weight of the last eggs on the bottom during storage causing them to crack.

Selection of hatching eggs.

The best eggs for hatching are those with a high level of shell gloss and a high level of pigmentation. Eggs with cracks or a high percentage of fragility when handled should be discarded, the latter being recognisable by a pale or whitish colour or simply a lack of shell at the time of collection, which is caused by a lack of calcium.

First 3 days.

During this time the incubator should not be opened at all as the embryo's development process is beginning.

Incubation procedure.

The incubator once opened on the third day should be left completely open so that the eggs are ventilated, remember that what is done is to replace the mother for an incubator and we must resemble as much as possible to the natural incubation process of a hen. To turn the quail eggs and if the incubator does not have a mechanism to turn the eggs automatically just put your hand on them very gently and begin to make circles to go turning them, this process can be done every 3 hours during the day or in the morning, at noon and in the afternoon.

Review of embryonic development.

To verify that the egg is formed and has been inseminated by the male on the 10th day, it can be checked by means of the ovoscope, the procedure is very simple, if the eggs in these ten days are clear to the translucent, it means that they were not fertilized, but if on the contrary the formation of blood endings is seen then that means the formation of the embryo.

Under normal incubation conditions there are two types of embryonic mortality; the first, important, in the course of the first days of development; and the second, less noticeable, in the last days.

Mortality in shell in the last stages of development and lengthening of the incubation period are the characteristic signs of not respecting the temperature and humidity at the time of incubation or hatching. The same applies to certain anomalies of the quail chick (shell stuck to down, splayed legs, crooked toes), the excessive number of which should alert the breeder. Lucotte G (p. 58)

Of hatching and birth.

From the time of candling until day 14 the embryonic development process is vital and therefore after this last day the eggs should not be touched until hatching begins. It should be noted that the incubator must have sufficient water in the tray to ensure the necessary humidity. At the end of the incubation process, a hatcher should be available to allow the newly hatched cotuplets to pass through to maintain their temperature and avoid contamination of the incubator.

Internal maintenance of the incubator.

In the incubator there must be total asepsis, this will prevent any bacteria from reducing or damaging the hatch. The use of creolin is recommended for disinfection after each hatch.

Unit 6

EGG PRODUCTION

The productivity of the female

The quail is a highly productive and economical animal in feed consumption, its production cycle is amazing, lasting up to a year from the first egg laid in week 7. The good performance is obtained until the first 6 or 7 months, after this period, the laying begins to drop a little, however many of the breeders leave them until it reaches a year and even more. The latter is not recommended, since it is not profitable for our commercial benefit to be feeding a bird without obtaining the main benefit which is the egg.

The lighting inside the shed.

To obtain a good result at the time of daily egg collection most breeders leave the bulbs on all night, however, this can also accelerate the deterioration of the bird because it does not have a relative period of rest. Excellent results can be obtained even without artificial light, these are switched off from 10 pm to 5 am and the performance is optimal.

Ambient temperature.

It is a very important part of egg production. Irregular temperatures lead to a decrease in daily laying, feather loss, etc. The ideal environment for quail is between 18º to 22º throughout the year without variations.

Good nutrition.

Feeding is another importantpart of quail breeding, in the adult animal as specified at the beginning of this book requires a high level of feeding, what is sought is that egg production is high. In this sense, all feeders should be kept with plenty of feed, otherwise, this would lead to the layers entering a state of stress or desperation to get the food and the percentage of daily laying would fall dramatically resulting in loss of money.

Egg collection.

This is a very delicate process, usually carried out daily and every morning by the staff that takes care of them. Near the end of the day the birds begin their laying cycle, so it is not recommended to have direct contact with them during this time. On the other hand, eggs collected in the morning hours should be deposited in special containers such as boxes with foam rubber bottoms or special boxes for quail eggs, this prevents their own weight from splintering the shell, which is extremely fragile.

The following images show the containers used to store the eggs safely. The material used is plastic, it has the ideal shape and size to guarantee their conservation.

20- Special case for storing quail eggs.

21- The cases are vertically viewed.

Unit 7

THE BREEDING OF THE COTUPOLLO

Data of the birth of the cotupollo.

Once the birth is done, the cotupolls must remain 24 hours in the incubator, after this, pass them to the hatcher, this procedure should not alarm people who are starting for the first time in the world of quail breeding because the newborn bird comes with a reserve called (*yolk*) capable of withstanding this period of time without food. The aim is for the quail to recover from the birth and at the same time for its plumage to slowly dry out. Finally, after hatching and after 24 hours, they must be hydrated with a solution (*serum*) before feeding them. This last point must not be ignored!

Similarly the cotupollo must receive a very good food if you want to get an accelerated growth, the first week is vital because in this period he is shedding his down and begins the process of change to the feathers. The food (starter) must be fully vitaminized because this bird in its growth stage is very delicate and any poor quality food leads to premature death, you must be very careful with this part.

Ways to raise the cotupollo.

There are two ways to raise the cotupollo, one is the homemade way, the other if you have a considerable breeding stock for large-scale marketing. In the first is very common to see people put the quail chicks to a small hen, this method is a little dangerous, almost always the hen does not recognize the cotupollos coming to cause death. The second method is the most recommended and is used in large-scale hatcheries. At the time of hatching, the cotupollo is placed in rooms with lighting or as I said before, the construction of a brooding hood that keeps the heat, should be circular. Moreover, the floor should be covered with rice husk to keep the quail dry and warm at all times.

23 - Cotupollos with 24 hours of life.

Birth defects.

Certain chicks are born with abnormalities leading to death. From experience, this is almost always genetic and comes from mixing of the same blood. There are other cases, for example, that lead to the quail's legs becoming stiff or simply twisting its head backwards, making it impossible to keep its balance.

Unit 8

QUAIL FOR HUMAN CONSUMPTION

The quail is one of the most profitable birds that exist, they occupy little space and leave a high economic profitability, not only for egg production and marketing, also for its meat, although its physical characteristics does not reach the weight of a broiler chicken, is intended for slaughter for its very particular flavor. In Europe and Asia they are a delicacy, for this reason people breed them with two purposes, for the production of eggs at commercial level and for fattening, in the latter the Coturnix Coturnix can weigh between 150 and 180 grams depending on their diet and demographic conditions, as well as the posture, the latter can be raised on the floor or in cages placed in the form of batteries, what would change is the type of food to use (*fattening ground vitaminized*).

> According to Lucotte (1976 p. 71), it is established that feeds of different compositions can be used for fattening, allowing, depending on the case, either to make an economy on the feed, or to accelerate the development of the animals, or both at the same time.

There are two ways to fatten a japonica quail, one of the most common is to put them on the ground like any other poultry, the other is in cages placed in batteries like layers. For fattening on the ground Lucotte 1976 stresses that ¨It allows the development of muscles and liveliness of the animals ... this type of breeding needs a very important surface.

On the other hand we have the broodstock in cages, it is the most suitableway to fatten the quail because less space is needed for their development. Lucotte 1976 points out that "as far as dimensions are concerned, the use of batteries is the same as that of breeders. The *roll away* device is certainly useless. Such a form of rearing allows a considerable economy of surface area in the premises.

Quail meat is very nutritious but in America there is no gastronomic culture with this bird, in these latitudes seems to be something very select and its consumption is reserved only in restaurants, unlike Europe where consumption *per capita* is high, in our continent is occasional.

Meat processing

Evisceration : the quail can be marketed whole or without the entrails, it will depend on the demand and the conditions between the producer and the distributor.

Preservation: after being slaughtered and prepared, they can remain in a place at low temperature such as a cellar or freezer, this will prevent their decomposition.

Packaging: for greater conservation and convenience, the quails can be placed in plastic containers when transporting them.

OVERALL RECOMMENDATIONS

For the breeding of japonica quail it is necessary to take into consideration many factors, although quail is a highly profitable animal in every way and resistant to diseases, the following must be taken into account:

1- Keep as much space as possible and well conditioned for the breeding of the bird.
2- Maintain hygiene in the facilities.
3- Provide good food for both Cotupollos and adult birds.
4- To build bells for the lifting of the Cotupollos.
5- Provide suitable feeders and drinkers for the quails.
6- In the case of the Cotupollos and at the time of power failure, you should not leave any drinking fountain inside the bell because they crowd around looking for a source of light and die drowned.
7- An electric generator should be available to provide heat to the incubator and in the hood.
8- Have an envelope of Osmolar Serum at the time of each birth, if you do not have it, prepare it, have a liter of water and add 3 tablespoons of salt with 1 ½ half tablespoons of sugar.
9- Turn the eggs in the incubator 3 times a day or more.
10- The trays should have sufficient water in order to maintain good humidity inside the incubator.
11- Split, small, calcium-deficient, misshapen eggs should be discarded when hatching.
12- The incubator must be disinfected at each hatch.
13- For a good disinfection use chlorine or bleach.
14- Always check the temperature in the incubators.
15- Avoid moving the reproductive females from their places, as this will cause a drop in posture.
16- Adult quails should have good space to move around the cage without difficulty, this prevents the spread of diseases.
17- The facility must contain good ventilation at all times, this will prevent ammonia build-up.

18-   The breeders should be placed in the water at least once every 15 days, vitamin supplement, one of the most used in Venezuela is *Multivit,* that applies to the chicks.

19-   The production area must be separate from the birthing room.

20-   There must be good lighting at all times.

21-   Care should be taken with the time of collection of fertile eggs, not to exceed the stipulated time.

22-   To find out if an egg is fertilized, it is necessary to use an ovoscope. It is recommended to check the eggs after 10 days of gestation.

23-   A good stock of feed should be available in case of insufficient supply. You should not resort to the purchase of feed made by hand, they lack the necessary nutritional supplements and vary in their chemical composition in each batch, they do not maintain the required standards.

24-   The quails that fulfill reproductive purposes should be rotated every six months since after this period they begin to lower the quality of their production. For commercial laying quails, the period can be extended to 1 year.

25-   Continuous washing of the drinking troughs will result in improved hygiene and the almost total elimination of disease.

BIBLIOGRAPHY

Rodrigo E. Vásquez R. & Higo H. Ballesteros C. **Farm Business Management. La Cría de Codornices (Coturnicultura)**" Produmedios 2007, Bogotá, Colombia.

G. Lucotte. The Quail Breeding and Exploitation. Ediciones Mund-Prensa 1985. Madrid, Spain.

M. Cumpa Gavidia. Breeding and Management of Quails. Practical Guide.

AUTHOR BIOGRAPHY

The author devotes much of his time to transmit his knowledge through writing, teaching and research. Graduated from the Arturo Michelena University, Valencia Edo. Carabobo, obtaining a degree in Modern Languages. He has a specialization in Teaching for Higher Education granted by the (University of Carabobo) in 2018. Specialization in Consecutive Interpretation from the Techno Lingüa Institute in 2016. Born in Caracas, Venezuela on March 3, 1982. He moved to Valencia in 1989, since then he has begun to capture his knowledge through books, he currently works as a freelance translator. He has been an English and French teacher at (Instituto Universitario Nuevas Profesiones) in 2013. (U.E. Fundación Valencia) in 2014-2015. (U.E. Valle Verde) in 2015-2016. (U.E. Jesús Manuel Berbín López) in 2015-2016. (U.E. San Miguel Febres Cordero) in 2017. (Unidad Educativa Batalla de Taguanes) in 2017. (Arturo Michelena University) in 2017-2018. He is the author of the following works: El Pozo del Vaquero and El Hotel Sangriento in 2018, Diez Postres Más Ricos I in 2018, Diez Postres Más Ricos II in 2019, Diez Postres Más Ricos III in 2020, Diez Postres Más Ricos IV in 2020, Diez Postres Más Ricos V in 2021, Niños Lobos in 2021.

Printed by Books on Demand GmbH, Norderstedt / Germany